Luciano Cavalcante de Jesus França
João Batista Lopes da Silva

Mapping areas vulnerable to pesticide contamination

Luciano Cavalcante de Jesus França
João Batista Lopes da Silva

Mapping areas vulnerable to pesticide contamination

Environmental management and land use planning using Geographic Information Systems (GIS)

ScienciaScripts

Imprint

Cover image: www.ingimage.com

This book is a translation from the original published under ISBN 978-3-330-75828-5.

Publisher:
Sciencia Scripts
is a trademark of
Dodo Books Indian Ocean Ltd. and OmniScriptum S.R.L publishing group

120 High Road, East Finchley, London, N2 9ED, United Kingdom
Str. Armeneasca 28/1, office 1, Chisinau MD-2012, Republic of Moldova, Europe
Printed at: see last page
ISBN: 978-620-8-29680-3

SUMMARY

ACKNOWLEDGMENTS

To God, family and science.

To CNPq (National Council for Scientific Development), for the research grant, which made part of my studies possible.

The Federal University of Piaui - UFPI, Campus Professora Cinobelina Elvas, for the opportunity to complete my academic training, and to be able to be part of its history and contribute to the viability of various projects and dreams.

My sincere thanks go to all those who have contributed in some way and given a little of themselves to make all this possible.

MY SINCERE THANKS!

SUMMARY

The Piauí savannah, which has satisfactory agricultural aptitude, has suffered a series of environmental impacts on natural resources, where intense exploitation for annual monoculture crops has changed the landscape and made it difficult to control and monitor land use and occupation. This makes it necessary and useful to use orbital data to monitor this change in native vegetation and other environmental degradation. With this in mind, the aim of this study was to determine the areas at risk of contamination by pesticides used in the production of annual monocultures in the Estiva creek basin, a tributary of the Uruçui Preto river, Piaui, based on a temporal survey of land use and occupation between 1984 and 2011. For this purpose, images obtained by the TM sensor of the Landsat 5 satellite were used to form mosaics in which the area of the Estiva stream basin is inserted. The images of the basin were used to highlight the vegetation using the Normalized Difference Vegetation Index (NDVI), in order to obtain the evolution of vegetation suppression in the basin, as well as maps of hydraulic conductivity, soil classes, slope, infiltration potential and the final map of the risk of pesticide contamination in the basin under study. The results show a high level of anthropization in the basin area, with an increase of around 55% in land use and occupation by agricultural crops, especially between 1984 and 2011, and an area of 269 km^2 , corresponding to 11% of the basin area, presenting a high potential risk of contamination by agrochemicals. Greater environmental attention is needed to better manage the use and conservation of these areas.

Keywords: Remote Sensing, Deforestation, Watershed Management.

1 INTRODUCTION

The term Cerrado is commonly used to designate the set of ecosystems (savannas, woodlands, fields and gallery forests) that occur in Central Brazil (RIBEIRO et al., 1981). It is Brazil's second largest biome, surpassed in area only by the Amazon. It occupies 21% of the country's territory and is considered to be the planet's last agricultural frontier (BORLAUG, 2002). However, this wealth is being intensively exploited by the livestock and agricultural sectors, which is supplying the high demand for food, but is threatening natural resources and biodiversity, not only because of the growing increase in cultivated areas, but often because of inadequate soil and vegetation management.

The way in which the plant, water and soil resources of the cerrados are used for agricultural purposes is the biggest cause for concern, based on the principle that the biome is characterized by fragile conditions, from the loss of biodiversity and vegetation to soil degradation. Given the sensitivity of the ecosystem, Cunha et al. (2008) point out that this biome has soils with low natural fertility, high acidity and low organic matter content (between 3 and 5%), as well as being subject to the seasonal climate. In order to cater mainly to the international market, the cerrado area has adopted the model of space occupation and production developed by *agribusiness* in the more developed countries, favoring large-scale, technology-intensive production, but neglecting environmental impacts.

Deforestation rates in the Cerrado have historically been higher than in the Amazon rainforest and the conservation effort in the biome is much lower than in the Amazon: only 2.2% of the Cerrado area is legally protected (KLINK and MACHADO, 2005). In view of this, it is necessary to promote conservation and sustainable use programs for this biome, with a view to promoting public policies and actions to conserve and preserve what remains of the native Cerrado forest.

Based on this scenario, after the Atlantic Forest, the Cerrado is the Brazilian biome that has suffered the most from human occupation, due to the growing pressure to open up new areas in order to increase the production of meat and grain for export, thus progressively depleting natural resources (BRASIL, 2009).

In a short space of time, the Cerrado areas have become one of the largest producers of soybeans, mainly by farmers from the south of the country and companies attracted by the low price of land and tax incentives granted by the government and the high price of soybeans on the international market (COSTA and SANTOS, 2010). In the state of Piaui, this reality has been characterized since the end of the 1980s, with growing participation in the Brazilian agribusiness scene. This region is characterized by intensive exploration of new areas opened up for agricultural crops, especially for soybean production. Piaui has a land area of 250,934 km^2 , around 11.5 million hectares of which are Cerrado, with a transitional area of 3.4 million hectares and an area suitable for cultivation of around 3 million hectares. These Cerrado areas are geographically located in different parts of the state, but predominate in the southwestern region and part of the extreme south of Piauí (REYDON and MONTEIRO, 2004).

Accelerated exploitation for the cultivation and production of annual crops in the cerrados has caused a number of problems in terms of the constant use of pesticides. These are characterized as useful for guaranteeing satisfactory productivity, but they have a major impact on the local environment. Barriuso et al. (1996) point out that the concern about this increase in the use of agrochemicals in grain production is mainly due to the fact that around 20% of the products used can reach water sources through rainwater runoff. In a study carried out in the state of Rio Grande do Sul, Bortoluzzi et al. (2006) found that the quality of surface water was contaminated due to the presence of active ingredients in some agrochemicals, and that the water in some streams bordering tobacco-growing areas was contaminated.

The southern region of the state of Piaui, characterized by being an ecotone (cerrado/caatinga transition), is known as the fastest growing agricultural frontier in the country. In view of this recognition, the impacts on natural resources and on communities in the regions surrounding these agricultural crops have been noted. The deforestation of native vegetation to incorporate new areas for the cultivation of soya and other annual monocultures, and the excessive use of pesticides by producers to combat diseases and pests, is a worrying reality. In this sense, there is a need for environmental work and surveys, assessment of water quality, aquatic and terrestrial biodiversity in areas close to those exploited by intensive and intensely mechanized agriculture, in order to add to and direct public policies for the conservation, preservation and sustainable management of the natural resources and soil of this biome.

Due to the accelerated dynamics of agricultural expansion in the southern region of Piaui, the consequent deforestation of cerrado areas and the inadequate management of pesticides and agrotoxins in general, cause environmental impacts. Thus, in the area of Geographic Information Systems and Remote Sensing, there is a need for studies that work on the characterization of this scenario, which has become increasingly anthropized, with areas that are increasingly cut up and heterogeneous.

The Riacho da Estiva basin, one of the largest tributaries of the Uruçui Preto river, is located in the southwestern region of the state of Piaui and passes through five municipalities: Uruçui, Sebastiâo Leal, Manoel Emidio, Palmeira do Piaui and Alvorada do Gurguéia. This basin has been suffering from the process of opening up new areas for grain production since the 1980s, as well as the increased use of pesticides to combat agricultural pests, which consequently causes a lot of damage and environmental impact. This decrease in natural resources, mainly due to anthropogenic causes, only adds to the concern for strategies aimed at conserving, restoring and managing these resources, which helps to ensure the stability of the ecosystem and the sustainability of the biome.

2. OBJECTIVES

2.1. General

Determine the areas at risk of contamination by agrochemicals in the Estiva Creek Hydrographic Basin, a tributary of the Uruçui Preto River, Piaui, Brazil.

3. THEORETICAL FRAMEWORK

3.1. Occupation of areas of the Brazilian cerrado by agricultural activities

The intensification of the occupation of the cerrado areas took place mainly from the 1980s onwards, due to government incentives for agricultural development programs and the low price of land, which was a decisive factor in the occupation of this biome, which, according to Moura (1997) began with reforestation of *Pinus* sp. and *Eucalyptus* sp., backed by Law 5.106, which granted tax incentives for reforestation. Subsequently, intensive agriculture was introduced with crops such as soya, cotton, coffee, corn, beans and peas, and the flat terrain allowed for heavy mechanization, rapidly changing the landscape by almost completely suppressing the natural vegetation cover.

The dominant technological model for agricultural exploitation in the Cerrados is deeply dependent on external inputs (calcium and fertilizers) produced by large chemical industries. This model, represented by the expansion of *commodities* and traditional planting techniques, is responsible for various environmental impacts on soils, water resources and biodiversity (RODRIGUES, 2005). In this way, the physical environment is intensely affected, with the degradation and depletion of springs and waterfalls, a reduction in the quality and quantity of water, and above all, pollution and chemical contamination of the soil and water, and consequent soil erosion and silting up of rivers. This is exacerbated by the adoption of inadequate management practices, which leads to soil degradation (FONTENELE et al., 2009).

The southern region of the state of Piaui, characterized by being an ecotone (cerrado/caatinga transition), is known as the fastest-growing agricultural frontier in the country. In view of this recognition, the impacts on natural resources and on communities in the regions close to these agricultural crops are noticeable. The deforestation of native vegetation to incorporate new areas for the cultivation of soya and other annual monocultures, and the excessive use of pesticides by producers to combat diseases and pests, is a worrying reality. In this sense, there is a need for environmental work and surveys, assessment of water quality, aquatic and terrestrial biodiversity in areas close to those exploited by intensive and intensely mechanized agriculture, in order to add to and direct public policies for the conservation, preservation and sustainable management of the natural resources and soil of this biome.

In the cerrado of Piauí, this occupation began in the 1980s, but it was only in the 1990s that agricultural production began in a significant way, through the installation of medium-sized entrepreneurs with the capacity to produce grain, particularly soybeans, through the use of modern techniques and inputs (GONÇALVES and MONTEIRO, 2007).

3.2. Environmental impacts of pesticide use on agricultural frontiers

Federal legislation defines the term pesticide in Law No. 7.802, of July 11,

1989, as:

Products and agents of physical, chemical or biological processes, intended for use in the production sectors, in the storage and processing of agricultural products, in pastures, in the protection of forests, native or implanted, and that of other ecosystems and also of urban, water and industrial environments, whose purpose is to alter the composition of flora and fauna, in order to preserve them from the harmful action of living beings considered harmful.

Table 1 shows that among the various agrochemicals, the most significant in terms of sales is herbicides.

Table 1 - List of the ten pesticides with the highest volume of consumption in the country in 2000.

Active Ingredient	**Class of use**	**National Consumption (t)**
Glyphosate	Herbicide	39.515,248
Sulphur	Acaricide	11.924,971
Atrazine	Herbicide	9.642,942
2,4-D (*Amine, ester, Triet, picloram*)	Herbicide	9. 016,003
Mineral oil	Acaricide, Adjuvant, Fungicide and Insecticide	8.618,376
Sulfosate	Herbicide	6.395,510
Mancozeb	Fungicide	5.434,920
Endosulfan	Insecticide	5.346,629
Copper Oxychloride	Fungicide	4.484,423

Trifluralin	Herbicide	3.313,580

Source: (IBAMA, 2000; PNGSQ, 2003)

The application of agrochemicals to soils and crops has become a common practice in agriculture, where accelerated exploitation has led to increased and constant use of pesticides. These are characterized as useful for satisfactory productivity, but when poorly managed and constantly applied in high doses, they cause major impacts on the soil, water and nearby ecosystems.

The emergence of chemical products used to combat crop pests represented great progress in agriculture, especially in large-scale production, bringing with it a new concept of production. The set of chemical products has been called pesticides, pesticides, pesticides, phytosanitary products or agrotoxins, the latter term being restricted to Brazil under Law No. 7.802/89 (KOTAKA and ZAMBRONE, 2001).

Faced with this scenario of the use of agrochemicals in the cultivated areas of the cerrados, the National Institute for the Processing of Empty Packaging (INPEV) is characterized. It works in conjunction with centers set up in agricultural hub cities to collect and dispose of agrochemical packaging, where 94% of the plastic packaging sold is disposed of in an environmentally correct manner. To this end, given the scenario of the use of chemical products and the consequent impact on the local environment, the various mechanisms for evaluating, surveying and analyzing the damage caused are shown, among various areas of study and research, including spatial geotechnologies.

Ramalho (2000) comments that the main objectives of using these agrochemicals are to increase the supply of nutrients (fertilizers), correct the pH of the soil (correctives) and protect crops by controlling diseases and pests (pesticides). These practices are efficient in controlling their respective problems, however, they can cause chemical degradation of the soil, as a result of toxic elements and/or compounds at undesirable levels. The same author also found in a study carried out to evaluate agrochemicals with heavy metals in the Caetés micro-basin, in Rio de Janeiro, that the sediments taken from the micro-basin stream showed heavy metal levels above the maximum standards established for drinking water by the Ministry of Health. He also found that the topography of the area and its agricultural use have a marked effect on the accumulation of heavy metals in the soil, i.e. the lower the slope and the more frequent the use of agrochemicals, the higher the content of heavy metals in the surface layers of the soil.

Marques (2005) also diagnosed the contamination caused by the use of pesticides in water bodies in a hydrographic basin belonging to the Ribeira de Iguape river, in São Paulo. He also concluded that water quality is influenced by seasonality, i.e. during rainy periods, a higher concentration of pesticides was detected in surface water caused by leaching from the soil by surface runoff.

In the environment, these products can act in two ways: by accumulating in the biota or by contaminating the water and soil, degrading the environment. This degradation of natural resources happens slowly and in some cases the damage caused by agrochemicals is irrecoverable (PORTO and SOARES, 2007). The dispersal of agrochemicals in the environment can also cause an ecological imbalance in the natural interaction between two or more species. Among this damage, we find the death of wild animals, insects, fish, water contamination, an increase in resistant pests and food contamination, with direct implications for human health (CASTRO, 1999). In view of this, the use of pesticides in the cerrado biome must be managed rationally in order to reduce the impact on the environment.

3.3. The new forest code and the expansion of deforestation in the Brazilian cerrado

The higher levels of deforestation in the Cerrado may be a reflection of the differences in the way the Forest Code treats the different Brazilian biomes: while only 35% of the area of agricultural establishments is required to be preserved as a legal reserve in the Cerrado, in the rainforest areas of the Amazon this percentage rises to 80% (Law 12.651/ of May 25, 2012).

Carvalho (2013) points out that the new version of the new Forest Code (Law 12.651/ of May 25, 2012) is still not what environmental protection and food production together actually require, but that it was already an advance on what remained in the version of the II CF (Law 4. 771/65) in relation to the original (Law 23.793/34).771/65) in relation to the original (Law 23.793/34), given that the changes were introduced during a period dominated by the dictatorship of the minority environmental movements. It is understood that progress in relation to environmental protection and food production concerns the intensity of inspections and the cohesive application of environmental policies, as well as food production on the premise of increasing its production and avoiding the clearing of new areas, promoting the recovery of areas that have already been degraded, and intensifying production in areas that have already been cultivated.

Article 12 of the New Forest Code deals with the delimitation of the Legal Reserve area, and shows that minimum percentages must be observed in relation to the area of the property, characterizing the legal reserve for properties located in cerrado areas as 35%.

In this context, the Forest Code (Law 4.771/1965) establishes a percentage of the rural property area as a Legal Reserve, which is necessary for the conservation of natural resources. In these areas determined as Legal Reserves, the vegetation cannot be cleared and can only be used on small properties under a sustainable forest management regime, using the area to obtain economic and social benefits in a planned manner, in accordance with established technical and scientific principles and criteria.

3.4. Geotechnologies as an ally in forest conservation

Geotechnologies are characterized as excellent allies for forestry work. Santos et al. (2012) reiterates that geotechnologies are applied to various forest resource factors, from the delimitation of ecological corridors, landscape ecology studies, environmental zoning, watershed management, forest fire risk, flood risk, environmental monitoring and adaptation, forest implementation projects and management plans for conservation units. In this scenario, the potential of geotechnologies to assess areas with intense land use and occupation and risks of pesticide contamination in watershed areas is appropriate.

In recent decades, orbital remote sensing has become an important tool for monitoring land resources, as it makes it possible to acquire significant data from large geographical areas. Remote sensing change detection studies are based on the fact that disturbances, both natural and man-made, in the environment produce detectable variations in the spectral response of the target or phenomenon being studied (ANDERSON, et al. 2003).

Remote sensing and georeferencing techniques have become important allies in the assessment of areas affected by natural and mainly anthropogenic disturbances, as well as demonstrating high efficiency in working with large areas, both because of the reliability of their consistency and because of their low cost. Lima and Zakia (2000) show that land use in a river basin has a considerable effect on the pulses of a river. In this context, environmental analysis of intensely fragmented landscapes with critical levels of remaining vegetation cover is fundamental, both for drawing up proposals to recover a river basin's capacity to provide ecosystem services and to prevent disorderly occupation and fragmentation of other landscapes.

It is known that remote sensing results in obtaining information about an object, surface or phenomenon through data analysis, without the device (sensor) coming into physical contact with it. According to Novo (2010), remote sensing can be defined as the joint use of sensors, data processing equipment placed on board aircraft, spacecraft or other platforms, with the aim of studying events, phenomena and processes that occur on the surface of planet Earth by recording and analyzing the interactions between electromagnetic radiation and the substances that make up its most diverse manifestations.

4. METHODOLOGY

4.1. Characterization of the Estiva Creek

The Estiva stream (Figure 1) is a tributary of the Uruçui Preto river (Figure 2), an important river in the south of Piaui, known for its dark waters due to the high concentration of organic matter and sediments, which cause turbidity in the river. The Estiva stream basin has a total drainage area of 3,099 km^2 and covers five municipalities, Uruçui, Sebastiâo Leal, Manoel Emidio, Palmeira do Piaui and Alvorada do Gurguéia, located in the west of the state of Piaui, between latitudes 7^0 28' S and 8^0 27' S and longitudes 44^0 17' W and 44^0 41' W (Figure 1).

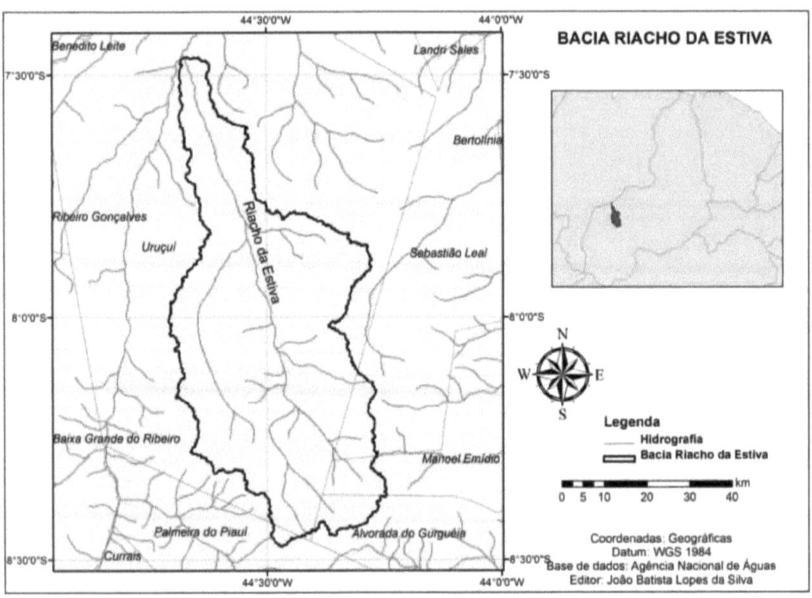

Figure 1 - Location of the Riacho da Estiva basin, a tributary of the Uruçui

Preto river, in the south of the state of Piaui.

Figure 2. stretch of the Uruçui Preto River, near the Estação Ecológica de Uruçui-Una (Photo: Luciano França).

Environmental information pianos were used to determine the trend of water behavior, infiltration and runoff, together with the transport of agrochemicals. Cross-referencing this information generated a map showing the areas at risk of contamination from the agrochemicals used (Figure 3). All these steps in this methodology were carried out using the GIS software ArcGIS10.

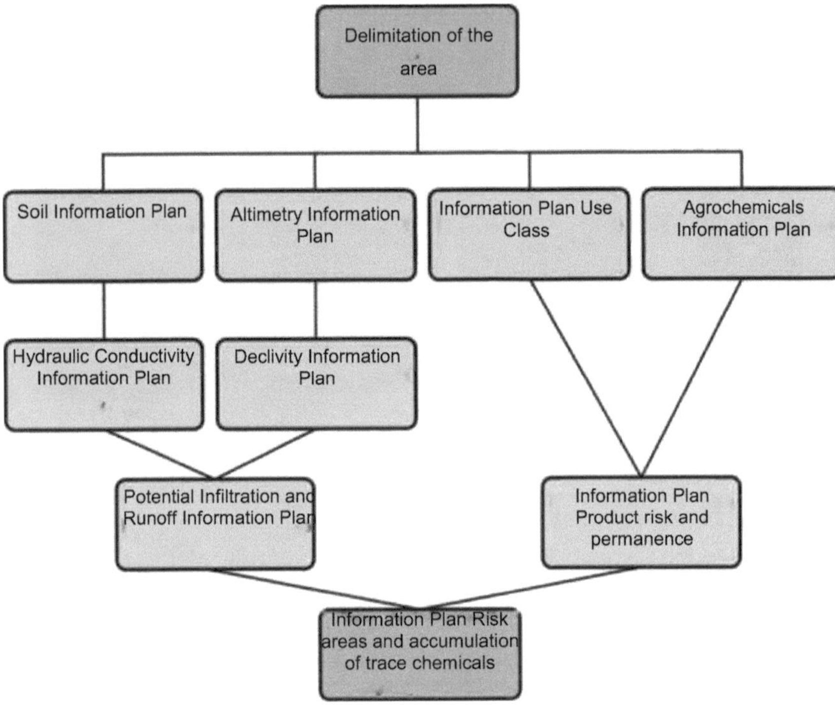

Figure 3: Flowchart of the crossing of information pianos for the production of the risk map. Adapted from Neves et al. (1998).

4.2. Information plans

For this stage, images were acquired from the Landsat (TM) 5 satellite on the INPE (National Institute for Space Research) website; SRTM raster images (USGS, 2005) containing elevation information (Digital Elevation Model) imported from the USGS (United States Geological Survey) website; images of the soil map on the IBGE (Brazilian Institute of Geography and Statistics) website; data on the quantity of agrochemicals used in recent years.

4.2.1. Altimetry and slope of the terrain

The altimetry information used came from the MDE (Digital Elevation Model) of the SRTM base (USGS, 2005), which, after removing the spurious depressions, deepening the channel and reconstructing the river banks, resulted in the MDHEC (Hydrologically Consistent Digital Elevation Model). The slope of the study area was obtained from the MDEHC and grouped into three classes: low, gentle and high, respectively containing slopes of less than 3% (low), between 3 and 8% (gentle) and between 8% and 20% (high). Areas with slopes above 20% are restricted to agricultural use.

4.2.2. Land Use Class

The information plan for the use class and subsequent intensity of land use and occupation was obtained by highlighting the vegetation using the Normalized Difference Vegetation Index (NDVI) from Landsat 5 TM images, bands 3 and 4, for the years 1984, 1987, 1990, 2003, 2005, 2008, 2009, 2010 and 2011.

To highlight vegetation, the Normalized Difference Vegetation Index (NDVI) was used, which takes into account the correlations between the reflectance of vegetation and soil, given that vegetation has a higher reflectance in the near infrared region and soil in the red region. In this case, the higher the density of the forest cover, the lower the reflectance in the visible (red) region and the higher the reflectance in the Near Infra Red (NIR) region. The NVI is obtained from the numerical expression:

$$IVDN = \frac{IVP - V}{IVP + V}$$

NDVI - Normalized Difference Vegetation Index

IVP - Near Infrared Region (Band 4)

V - Red Region (Band 3)

After highlighting the vegetation by IVDN, the images were classified by non-supervised classification into two classes: native vegetation and anthropized areas. In order to check which areas of the basin were most exploited and, consequently, had the greatest traces and residues of agrochemicals, the classified IVDN images from all the years were added together.

4.2.3. Soils

The classification of the soil map (soil chart) from the IBGE website was used for the soil type information plan. For the qualitative determination of hydraulic conductivity, the variables texture, structure, aggregate stability and soil depth were taken into account. The combination of these four variables resulted in three groups of hydraulic conductivity: high, medium and low. The crossing of the two information planes, hydraulic conductivity and slope classes, gave rise to a third information plane called infiltration potential and surface runoff (tables 2 and 3).

4.2.4. Pesticides

In order to gain an understanding of the products used in the cerrados, a survey was carried out of which products have been used in the region in recent years. This data was obtained from SEMAR (State Secretariat for the Environment and Water Resources) in the municipality of Bom Jesus-PI.

4.3. Final risk map

The final analysis was carried out considering the places where factors that aggravate the risk occurred simultaneously, which generated a final map indicating the areas that deserve greater attention. The accumulation of surface runoff indicates the greatest concentration of agrochemical traces within the basin.

Table 2. Hydraulic conductivity classification of the main Brazilian soil classes, at a high categorical level, according to some of the soil's physical properties.

Soil Class	Texture Dominant	Structure Dominant	Aggregate stability	Depth (Horiz. A+B)	Conductivity Hydraulics
Latosol Purple	Clay	Angular to sub-angular blocks	Stable	Deep	High
Latosol Red Dark	Clay	Sub-angular blocks	Stable	Deep	High
	Clay Sandy	Granular	Little Stable	Deep	Average

	Sandy	Granular	Little stable	Deep	Average
Latosol Red Yellow	Clay	Sub-angular blocks	Stable	Deep	High
	Clay Sandy	Granular	Little Stable	Deep	Average
	Sandy	Granular	Unstable	Deep	Average
Latosol Yellow	Clay	Sub-angular blocks	Stable	Deep	Average*
	Clay Sandy	Sub-angular blocks	Stable	Deep	Average*
	Sandy	Sub-angular blocks	Stable	Deep	Average*
Terra Roxa Structured	Clay	Blocks Angles	Stable	Deep	High
Podzolic Red Dark	Clay	Sub-angular blocks	Stable	Relatively Deep	Average
	Clay	Sub-angular blocks	Stable	Relatively Deep	Average
Podzolic	Clay	Sub-	Stable	Relatively	Average

Red Yellow		angular blocks		Deep	
Podzolic Yellow	Clay	Sub- angular blocks	Stable	Relatively Deep	Average

Source: Gomes et al. (1996).

Table 3. Infiltration potential based on hydraulic conductivity and slope.

Conductivity	Slope		
	Low (<3%)	Mild (3 to 8%)	High (8 to 20%)
Low	Medium	Bass	Bass
Average	High	Medium	Bass
High	High	High	Medium

Source: Neves et al. (1998).

5. RESULTS AND DISCUSSION

5.1. Normalized Difference Vegetation Index (NDVI)

The IVDN results from the Landsat 5 TM images for the years 1984, 1987, 1999, 1993, 1996, 1999, 2000, 2005, 2008, 2009, 2010 and 2011 show strong anthropization throughout the basin, especially in the areas of native vegetation of the cerrado ecosystem, resulting in an increase in agricultural areas (Figure 4).

With the results of the NDVI images (Figure 4), it can be seen that there has been an evolution in the anthropization of the area corresponding to the Estiva stream basin, corroborating Reydon and Monteiro (2000), who state that the occupation of the Piauí cerrado began in the 1980s, but with effective productivity only in the 1990s, through large-scale production, mainly of soya. A comparison between 1984 and 2011 shows a rapid expansion of agricultural activities and a reduction in native vegetation, evidenced by the areas in red, which characterize farms and agricultural plots.

There was a noticeable increase in the number of farms in the watershed area between 1984 and 2011, showing that farming began in the area at the beginning of the 1980s, but that it became more noticeable from 2000 onwards. It can be seen that between the 1990s there was a considerable presence of fires, which is characterized by the undefined patches in the 1990 image, differentiating them from the regular polygons of farms and agricultural plots, as shown in Figure 4.

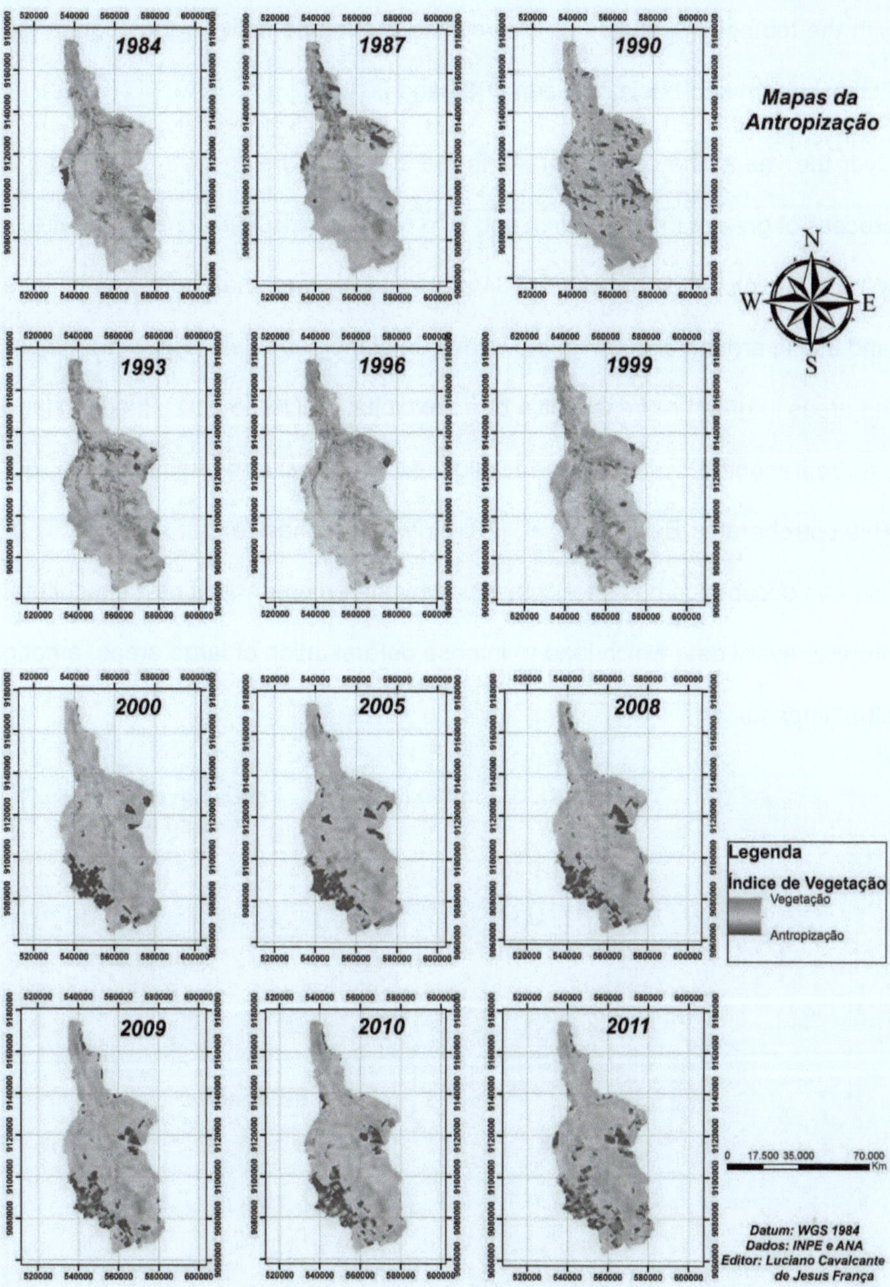

Figure 4 - Evolution of the Normalized Difference Vegetation Index (NDVI), with the temporal evolution of anthropized areas and native vegetation in the Estiva stream sub-basin between 1984 and 2011.

Over the years, the cerrado areas in the sub-basin have been undergoing a process of greater intensification and land use for agricultural crops. The areas with high exploitation between 1984 and 2011 showed that there was intense land use in around 269 km^2 of the total area of the sub-basin (Figure 5). Where the areas in green represent little or no exploitation (1,685 km2), medium land use, represented by the areas in yellow, (1,144 km2) and intense use in red. This corroborates Barbosa et al. (2012), who emphasize that the process of use and occupation has been intensified by the emergence of new agricultural projects every day, which lead to intense deforestation of large areas, among other impacts.

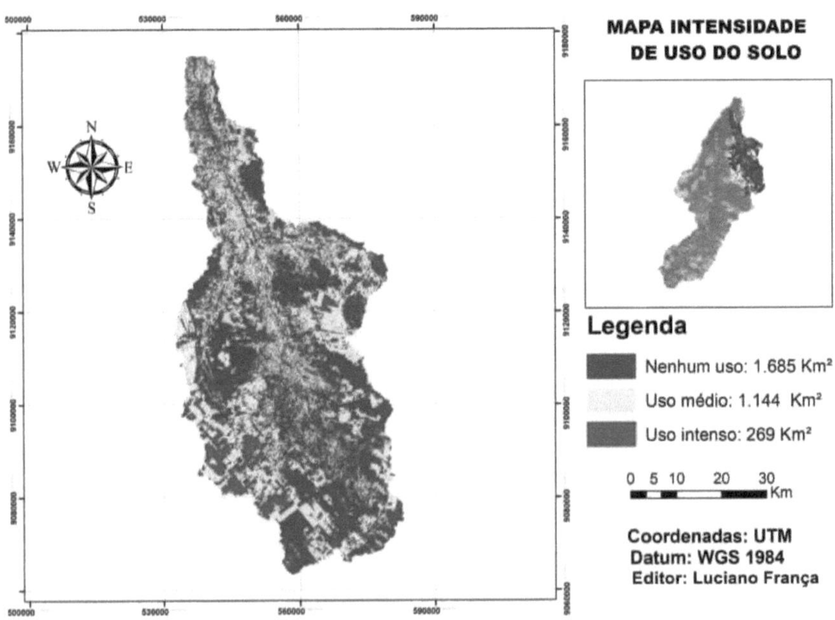

Figure 5 - Map of soil exploitation intensity between 1984 and 2011, resulting from the sum of classified IVDN images, in which the areas in green show little or no exploitation, in yellow medium use, while the areas in red show several years of exploitation and use.

The results in numbers of the evolution of deforestation and loss of native vegetation in the hydrographic basin under study are shown in Table 4.

Table 4 - Evolution of areas of agricultural occupation and anthropization compared to 1984 and 2011

Year	Native vegetation	Anthropization
1984	2,761 km^2	337 km2
1987	2,428 km2	670 km2
1990	2,345 km2	754 km2
1993	2,506 km2	593 km2
1996	2,542 km2	557 km2
1999	2,539 km2	560 km2
2000	2,510 km2	589 km2
2005	2,508 km2	591 km2
2008	2,473 km2	626 km2
2009	2,512 km2	587 km2
2010	2,430 km2	668 km2
2011	2,321 km2	778 km2

From the results obtained, according to Table 4, in this 27-year period there has been a significant increase in anthropization and a consequent decrease in the basin's forest vegetation.

5.2. Terrain slope classes

It was possible to obtain the drainage area of the Estiva stream sub-basin from the MDEHC, however, in order to arrive at the perfected MDEHC, it was necessary to remove the spurious depressions responsible for preventing progressive surface runoff at certain points in the basin. Using the MDEHC, it was possible to determine the slope classes of the terrain through the MDE (Figure 6).

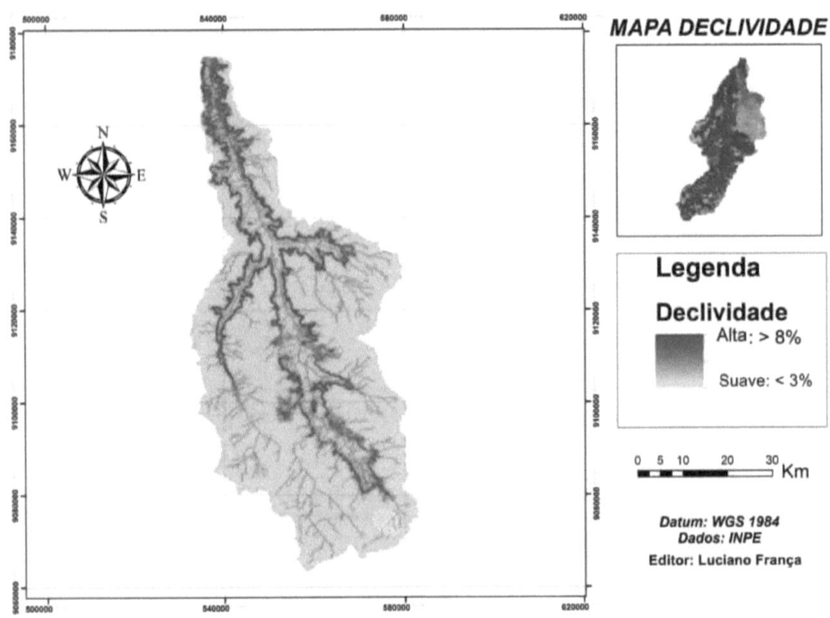

Figure 6. Slope classes of the Estiva stream sub-basin, with low slope in yellow and high slope in red.

It was found that most of the sub-basin had a low slope, shown in yellow, a slope of less than 3%, which is suitable for the use of agricultural machinery and implements used in planting, harvesting grain and other agricultural activities, a scenario suitable for the intense mechanization of production (Figure 7). The Cerrado latosols in the Midwest occupy 90 million hectares - the most favorable area for grain crops: the soils are deep, well-drained, with slopes usually less than 3% MAROUELLI (2003).

5.3. Soil map and hydraulic conductivity

The Yellow Latosol soil class is predominant in the basin, corresponding to the main crop areas, followed by Red Latosol and a small portion of Neosols (Figure 7). Knowledge of soil typology is an important factor, since the soil's physical properties and attributes are valid for determining the soil's hydraulic conductivity (Figure 8). Latosols predominate in the Brazilian Cerrado and, according to Santos et. al. (2012), are present in 46% of the Cerrado biome area. They are characterized as soils with low fertility and high acidity; above all, they are old, deep soils with good drainage and are located on flat or slightly undulating terrain.

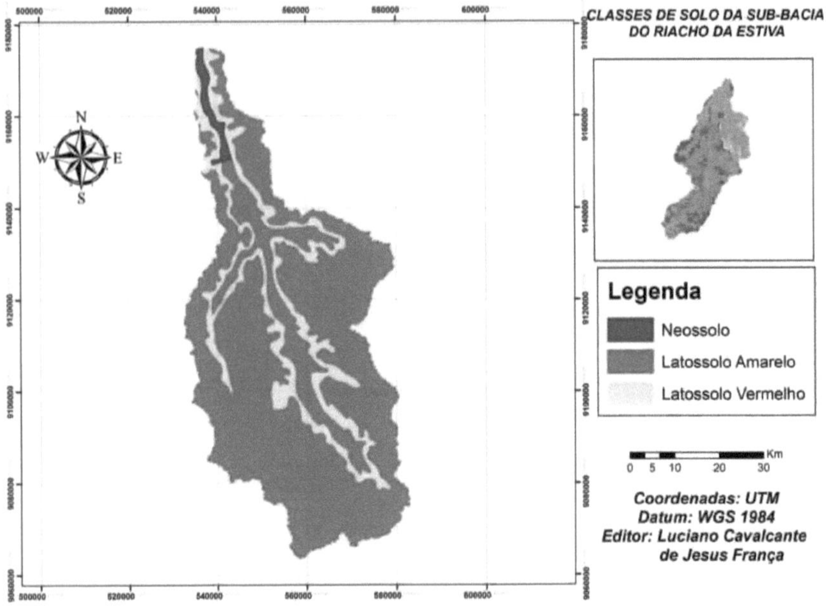

Figure 7 - Map of soil classes in the Estiva stream sub-basin.

The hydraulic conductivity map (Figure 9) was also drawn up based on the soil classes. It can be seen that most of the basin has low hydraulic conductivity, which encourages the accumulation of water and consequently the permanence of pesticide traces. It can also be seen that the areas with the highest water conductivity are in the lowlands, which are uncultivated areas. These areas often receive a lot of sediment from rainfall runoff, accompanied by traces of pesticides, reaching the native forest areas and local communities.

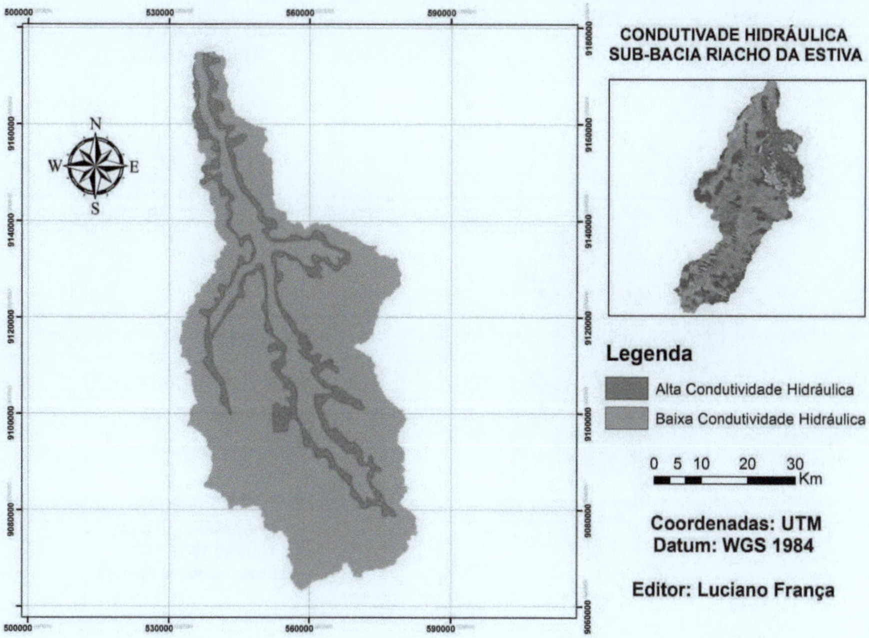

Figure 8. Hydraulic conductivity of the Estiva Creek sub-basin.

5.4. Maps of Infiltration Potential and Surface Runoff

By cross-referencing the soil slope (Figure 6) and hydraulic conductivity (Figure 8) information plans, a map of the potential for water infiltration and soil surface runoff was generated (Figure 9). Different soil infiltration potentials can be seen. The sites with high infiltration potential were the same sites with low soil slopes, demonstrating the high infiltration capacity of the soil in these areas. In the areas where large agricultural projects are located, the soil has a high infiltration potential, which could lead to groundwater contamination.

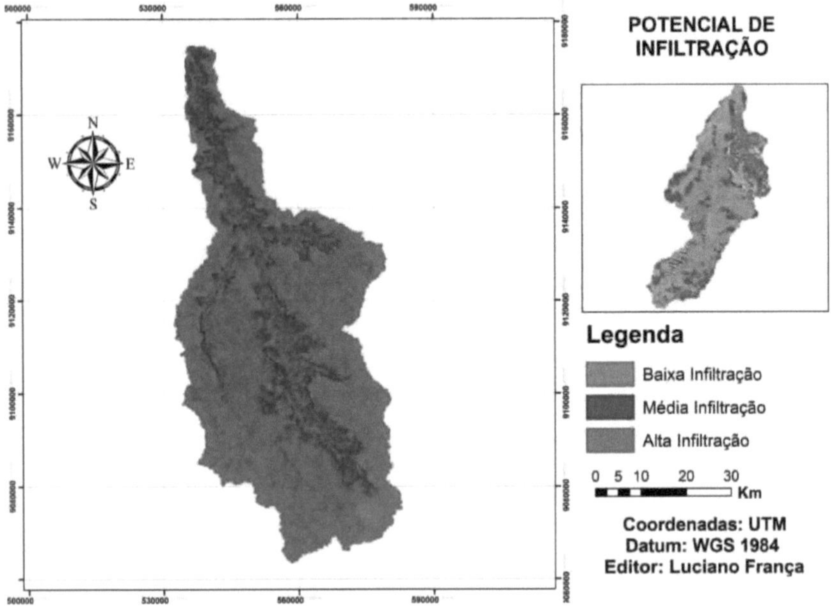

Figure 9. Infiltration potential and surface runoff in the Estiva stream sub-basin.

5.5. Final contamination risk map

To obtain the final map of pesticide contamination risk, the infiltration potential and land use were cross-referenced (Figure 10). The areas with gentle or slightly undulating slopes were those with the highest concentrations of agricultural properties, possibly growing soybeans and other monocultures, resulting in the frequent use of agrochemicals. These products can percolate or be leached into the soil profile due to their high infiltration potential and consequently cause environmental problems with groundwater contamination. Ribeiro and Vieira (2010) state that water is one of the primary routes by which pesticides are transported from the places where they were applied to other compartments of the hydrological cycle, and that contaminants can reach surface water through rainwater runoff and irrigation, or underground water through drainage and percolation into the soil. Therefore, there is a possibility of contamination of areas that have properties close to surface water sources through surface runoff, which could cause serious environmental problems resulting from the increasing use of pesticides to control pests and diseases, a common practice in conventional agriculture established in the Estiva stream sub-basin to ensure satisfactory grain production.

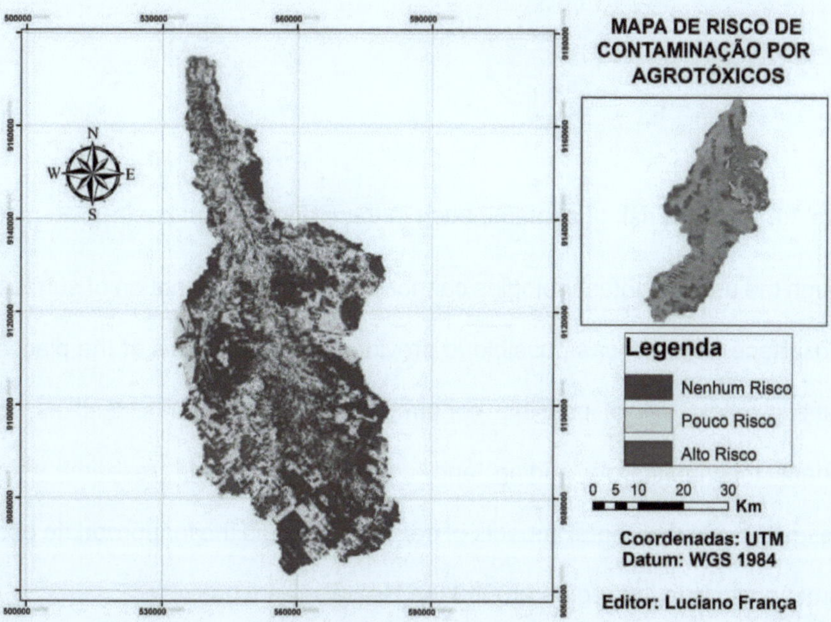

Figure 10. Final map of pesticide contamination risk in the Estiva stream sub-basin.

It can be seen that areas with a high potential risk of pesticide contamination account for around 11% of the total area, corresponding to 269 km^2 ; areas with a low or medium risk of contamination account for 44% of the total area, corresponding to 1,144 km^2 ; and areas with no risk account for 45% of the total area, 1,685 km^2 as shown in Figure 11. The quantification of the area is obtained from the equationalizer in the ArcGis *software.*

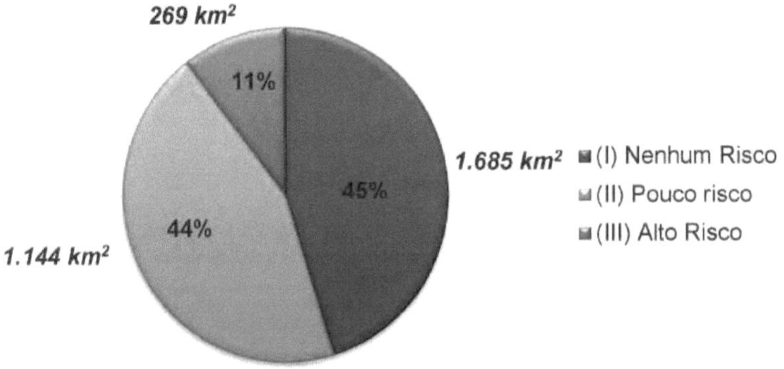

Figure 11. Areas at risk of contamination in the Estiva stream basin.

Although the use of geotechnologies cannot show the contamination of surface or subsurface water, it was possible to provide a concrete basis of the places with the greatest risk of possible contamination. This presents itself as an important mechanism for aiding and complementing field research that assesses the environmental impacts of deforestation and the inappropriate use of agrochemicals in the region around the Estiva stream basin.

Table 5 shows the data on the toxicological classes of agrochemicals used on farms in the sub-basin.

Table 05. Percentage and number of products in all classes of pesticide toxicity used in the region of southern Piaui over four years (State Secretariat for the Environment and Water Resources - SEMAR, 2013).

		YEARS							
		2009		**2010**		**2011**		**2012**	
		N° products	(%)	N° products	(%)	N° products	(%)	N° products	(%)
Toxicity classes	**I**	108	22,13	102	21,16	251	24,95	118	21,03
	II	102	20,9	124	25,73	229	22,76	116	20,68
	III	199	40,78	154	31,95	349	34,69	250	44,56
IV		79	16,19	102	21,16	177	17,59	77	13,73
TOTAL		**488**	**100**	**482**	**100**	**1006**	**100**	**561**	**100**

Where: class I - extremely toxic; class II - highly toxic; class III - moderately toxic; class IV - slightly toxic.

It can be seen that in 2012, compared to 2009, there was an increase in the use of pesticides in the extremely toxic, highly toxic and moderately toxic classes, with only the low toxic class showing a slight reduction in use. This shows a considerable increase in a comparison of only 4 years, which requires greater attention, monitoring of sales of these products and greater concern about the possible environmental impacts caused by this increase in the use of agrochemicals to combat pests and diseases.

6. CONCLUSIONS

The results obtained in this research indicated that the GIS tools used contributed to the generation of the maps and the concrete interpretation of their features. In the Estiva stream sub-basin, based on average and intense land use, there was an increase of around 55% in land use and occupation by agricultural crops and a consequent increase in the deforestation of native forest areas in the basin between 1984 and 2011; and an area of 269 km^2 , 11% of the basin area, presents a high potential risk of contamination by agrochemicals, while 1.144 km^2 (44%) present little risk of contamination and 1,685 km^2 (45%) present no risk of contamination by agrochemicals. These results provide fundamental support for the development of cerrado conservation programs and for the proper management of the activities that take place within this hydrographic basin.

7. BIBLIOGRAPHICAL REFERENCES

ANDERSON, O. L. ROJAS, E. H. M.; SHIMABUKURO, Y. E. **Advance of Soy on Cerrado and Forest Ecosystems in the State of Mato Grosso.** Proceedings of the XI SBSR, Belo Horizonte, INPE, 2003.

BARRIUSO, E.; CALVET, R.; SCHIAVON, M.; SOULAS, G. **Les pesticides et les pollluants organiques des sols: transformation set dissipation**. Étudeet Gestion des Sols, Ardon, v.3, n.4, p.279-296, 1996.

BORLAUG, N.E. **Feeding a world of 10 billion people: the miracle ahead**. In: R. Bailey (ed.). Global warming and other eco-myths. pp. 29-60. Competitive Enterprise Institute, Roseville, USA, 2002.

CARVALHO, L. A. de. **The new forest code commented on: article by article, with changes brought in by Law 12.727, of 17.10.2012 and references to Decree 7.830, of 17.10.2012.** Curitiba: Juruà, 584p, 2013.

CASTRO, M. S. J. **Agrotoxics Use Practices in the Municipality of Cachoeira Macacu. RJ: An Environmental Health Study.** 1999. 159p. Dissertation (Post-Graduation in Environmental Science) - Center for General Studies, Fluminense Federal University. Niterói. 1999.

COSTA, A. R; SANTOS, O. F.Expansão Agricola e Vulnerabilidade Natural do Meio Fisico no Sul GoianoOepartamento de Geografia da FCT/UNESP, Presidente Prudente. **Revista Geografia Em Atos**. n°. 10, vol, 2. p. 23-35, 2010.

CUNHA, N. R. da S.; LIMA, J. E. de L.; GOMES, M. F. de M.; BRAGA, M. J. **The Intensity of Agricultural Exploitation as an Indicator of Environmental Degradation in the Cerrado Region, Brazil.** RER, Piracicaba, SP. Vol. 46, n° 02, p. 291-323, 2008.

EMBRAPA. **Legal reserves in the Cerrado Biome: use and preservation.** Documentos, 158, Planaltina, DF: Embrapa Cerrados, 2006.

FONTENELE, W.; SALVIANO, A. A. C.; MOUSINHO, F. E. P. Physical attributes of a Yellow Latosol under management systems in the cerrado of Piauí. **Revista Ciência Agronômica**, vol.40, p.194-202, 2009.

KOTAKA, E. T.; ZAMBRONE, F. A. D. Contributions to the Construction of Guidelines for Assessing the Toxicological Risk of Pesticides. Campinas, SP: ilsi, 2001. 160p.

LIMA, W. P.; ZAKIA, M. J. B. **Hydrology of riparian forests**. In:. RODRIGUES & LEITAO FILHO (ed.). **Riparian Forests - Conservation and Recovery**. Sâo Paulo: EDUSP/FAPESP: 2000. p. 33-44.

MARQUES, M. N. Evaluation of **the impact of pesticides in environmental protection areas, belonging to the hydrographic basin of the Ribeira de Iguape River, Sao Paulo, a contribution to the critical analysis of the legislation on the potability standard**. Doctoral thesis. Nuclear and Energy Research Institute, USP, São Paulo, 2005.

MOURA, V. P. G. Research with Eucalypitus and Pinus in the cerrado region.

In: SYMPOSIUM ON THE CERRADO VII: Utilization Strategies, 1989, Brasilia. **Proceedings** ... 2.ed. rev. Planaltina: EMBRAPA - CPAC, p.183-197,1997.

NEVES, M. C.; GOMES, M. A.; LUIZ, A. J. B.; Spadotto, C. A. Geographical information systems. Applications in agriculture. In: ASSAD, E.D. SIG na avaliaçâo do impacto ambiental por agroquimicos. 2ª Ed. Revista e ampliada, Brasilia. **EMBRAPA**, 1998, p.241-250.

NEW. M. L. E; **Remote Sensing: Principles and Applications.** Sâo Paulo: Blucher, 2010. 387p.

GONÇALVES, G. L; MONTEIRO, L. S. M. O. Preliminary Environmental Impact Study and the Environmental Adequacy of Agricultural Enterprises in the Cerrados of Piaui. In: "VII Meeting of the Brazilian Society of Ecological Economics". **Proceedings...** Fortaleza, 2007.24p.

KLINK, C. A.; MACHADO, R. B. **A conservaçao do cerrado brasileiro.** Megadiversidade. v, 1. n, 1. p. 148 - 155. July, 2005.

MAROUELLI, R. P. o. **Desenvolvimento Sustentâvel Da Agricultura No Cerrado Brasileiro**.Brasilia: ISAEFGV/ Ecobusiness School, 54p. 2003.

RAMALHO, J. F. G. P.; SOBRINHO, N. M. B. A.; VELLOSO, C. X. **Contamination of the Caetés watershed with heavy metals due to the use of agrochemicals.** Pesquisa Agropecuâria Brasileira, Brasilia, v.35, n.7, p. 1289-1303, jul. 2000.

REYDON, P. B., MONTEIRO, L. S. M., **Occupation of the Piauí Cerrado. A**

Process of Land Valuation, 2000.

REYDON, P. B., MONTEIRO, L. S. M., **Occupation of the Piaui Cerrado. A Process of Land Valuation**. 2004.

RIBEIRO, J.F.; SANO, S.M.; SILVA, J.A. **Chave preliminar de identificaçâo dos tipos fisionômicos da vegetaçâo do Cerrado**. p. 124-133 In: Anais do XXXII Congresso Nacional de Botânica. Botanical Society of Brazil, Teresina, 1981.

RODRIGUES & LEITAO FILHO (ed.). **Riparian Forests - Conservation and Recovery**. Sâo Paulo: EDUSP/FAPESP: 2000. p. 33-44.

RODRIGUES, W. **Economic Valuation of the Environmental Impacts of Planting Technologies in the Cerrado Region.** RER. Vol. 43, n° 01, p. 136-153, 2005.

SANTOS, A. R.; PELUZIO, J. B.E.; PELUZIO, T. M. O.; SANTOS, G. M. A. D. A. **Geotechnologies applied to forest resources.** CAUFES. 249 p. Alegre, ES: 2012.

SOARES, L.W; PORTO, F.M; Agricultural Activity and Environmental Externality: An Analysis of the Use of Agrotoxics in the Brazilian Cerrado. **Journal Ciência e Saùde Coletiva.** p. 131-143. 2007.

MIX
Papier aus verantwortungsvollen Quellen
Paper from responsible sources
FSC® C105338

Printed by Books on Demand GmbH, Norderstedt / Germany